LEARNING OJIBWE WITH

NUMBERS & ANIMALS

GAA-IZHI-AZHENAADIZIYANG NINDINIMAAGANINAAN: CONNECTING WITH OUR FIRST FAMILY

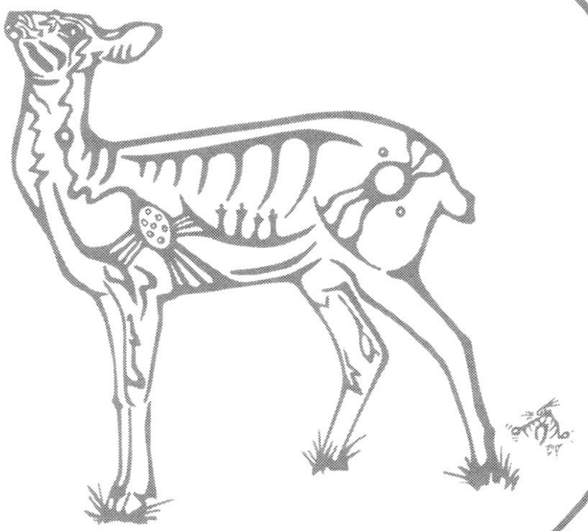

Written by Nicole Ineese-Nash
Images by Nyle Miigizi Johnston

TakingITGlobal
INSPIRE INFORM INVOLVE
Connected North

DEDICATED TO

Shreya, Sage and Xavier

Author's Statement:

Connection is at the core of what it means to be human. As Anishinaabekwe, I have been taught this in a multitude of ways. I am humbled to have been given the space to write and reflect in this book about what our relations to place and the beings therein mean to our own lives and our collective wellbeing. I am also inspired to consider what our responsibilities may be in honouring these relationships in the context of settler-colonialism and environmental change.

I carry the name of Songe Winnishe Wabigwanikwe, or Strong Beautiful Wildflower woman. In carrying this name I recognize my responsibility to walk this earth in a manner that holds all the gifts of creation in high regard.

Wildflowers may grow when no one is tending to them, but they thrive because everything contributes towards their growth. There are webs of relations that extend beyond our conception and these relationships help us all to live. For me, connection is lifegiving; it is spirit nourishing; it is everything. The path I walk is a one of rediscovery; of who I am, where I come from, and what my purpose is. It is through learning about my family (both in the human and non-human sense), my culture, and the land of my ancestors that I have come to know myself.

My lineage comes from the place where three rivers meet, or Mamawmattawa, near Constance Lake First Nation in Treaty 9 territory where I am a registered band member. Growing up in Tkaronto, I have longed for a connection to something I did not quite know I was missing. In working with Indigenous children, youth, and communities over the years, I have realized this disconnect is felt by many. I am now an auntie, raising two young children, and I wish for them to grow as a garden sustained by love and fed by wisdom. I wish for them to never question where they fit within the world, and I see education as a means to break the cycles of disconnect that have been perpetuated in my bloodline through colonial trauma.

I am a doctoral student at the University of Toronto, studying Social Justice Education and Indigenous Health. I also hold a Master's degree in Early Childhood Studies. This educational background has given me a passion for education for children and youth of all ages in manners that are inclusive and rooted in cultural knowledge. I hope this book provides an entry point into the conversation about Indigenous worldview in the classroom and a mechanism for all children to share in the gifts Indigenous knowledge can provide.

Miigwetch

Nicole Ineese-Nash

Artist Introduction

It is easy to forget that we are all children of Our First Mother of Creation. Examples can be found throughout nature that connects us to our First Family and Plant Nation relatives. From the flowing river systems that are our veins, to the cedar tree that is our brain stem, the patterns of connectivity are not only beautiful but also inspiring. When I draw or paint I humble myself and try to capture a fragment of the messages, stories, instructions, teachings and beauty that Creation has to provide.

My spirit name Wiishkoonseh Miigizi'enh which means Whistling White Headed Eagle. I come from a Storytelling background and I have apprenticed with Storytellers since I was young. I have a responsibility to continue this legacy for the healing of my own people while connecting to the broader world. Sources of my artistic inspiration include woodland painters, Story Tellers and the traditions of my culture.

In a time of reconciliation, it is important for all people to know that we exist and have such a strong, beautiful legacy of stories and teachings. Our stories resonate with people across different cultures and they carry universal messages of love, kindness, fairness and care for Mother Earth. We each have a responsibility to share our gifts and I have the gift of being a Story Teller.

My logo is my signature and represents who I am. Talking about Spirit names is talking about tradition and culture. When I am asked about my art, the stories come out and people become captivated and interested in learning more. It creates more curiosity from people in learning about the history of our people. It is planting seeds in the minds of people so they can become more aware. Together in partnership with TakingITGlobal, I developed the Connecting With Our First Family collection of 81 images of animals and plant medicine that have been featured as part of Ojibwe Language learning resource materials and I feel grateful to be able to share the visual stories with you as part of this children's book.

Miigwech

Nyle Miigizi Johnston,
Chippewas of Nawash Unceded First Nation

Community Partner Introduction

Art has a way of provoking our senses and inviting us to explore different ways of understanding ourselves and the world around us. Images can capture our imagination and draw us into a moment, an emotion, a teaching and into a deeper state of curiosity.

Over the past two years, TakingITGlobal's Connected North program has partnered with Indigenous Artist and Visual Story Teller Nyle Miigizi Johnston to develop an Ojibwe Language and Education Resource called Connecting with our First Family. The collection of images featuring different line drawings of animals, birds and insects created by Miigizi are being transformed into a series of educational resources, games and learning activities for classrooms and youth groups. We are delighted to collaborate with Author Nicole Ineese-Nash on this children's book to support learning numbers and animals in Ojibwe.

This growing body of work contributes to supporting support students and educators in the process of understanding and appreciating the many diverse Indigenous languages, cultures and communities while strengthening connection to identity and our relationship to the land. The project also serves as a catalyst for developing resources in other Indigenous languages in communities where Connected North partner schools are based.

We would like to acknowledge that there are many different dialects and written forms across the diverse Indigenous languages that have been translated as part of this resource and we are grateful to the community members who have shared their teachings, understanding and translations as part of this project. This book has been translated by Jason Jones from Treaty 3 who has been working with our Connected North team in partnership with Rainy River District School Board. We are happy to be able to share this learning experience with you and hope it will inspire children, youth to develop their own ideas, creativity and talents as leaders of their communities who have a role in shaping the future.

Niá:wen

Merci

Quanaqutin

Ci rishiwihk ᒥᑵᐨ

Miigwech L'o Mikwec

Gunalchéesh

Wela'lioq

Thank you

Mársi

Jennifer Corriero

Executive Director,

TakingITGlobal and Connected North

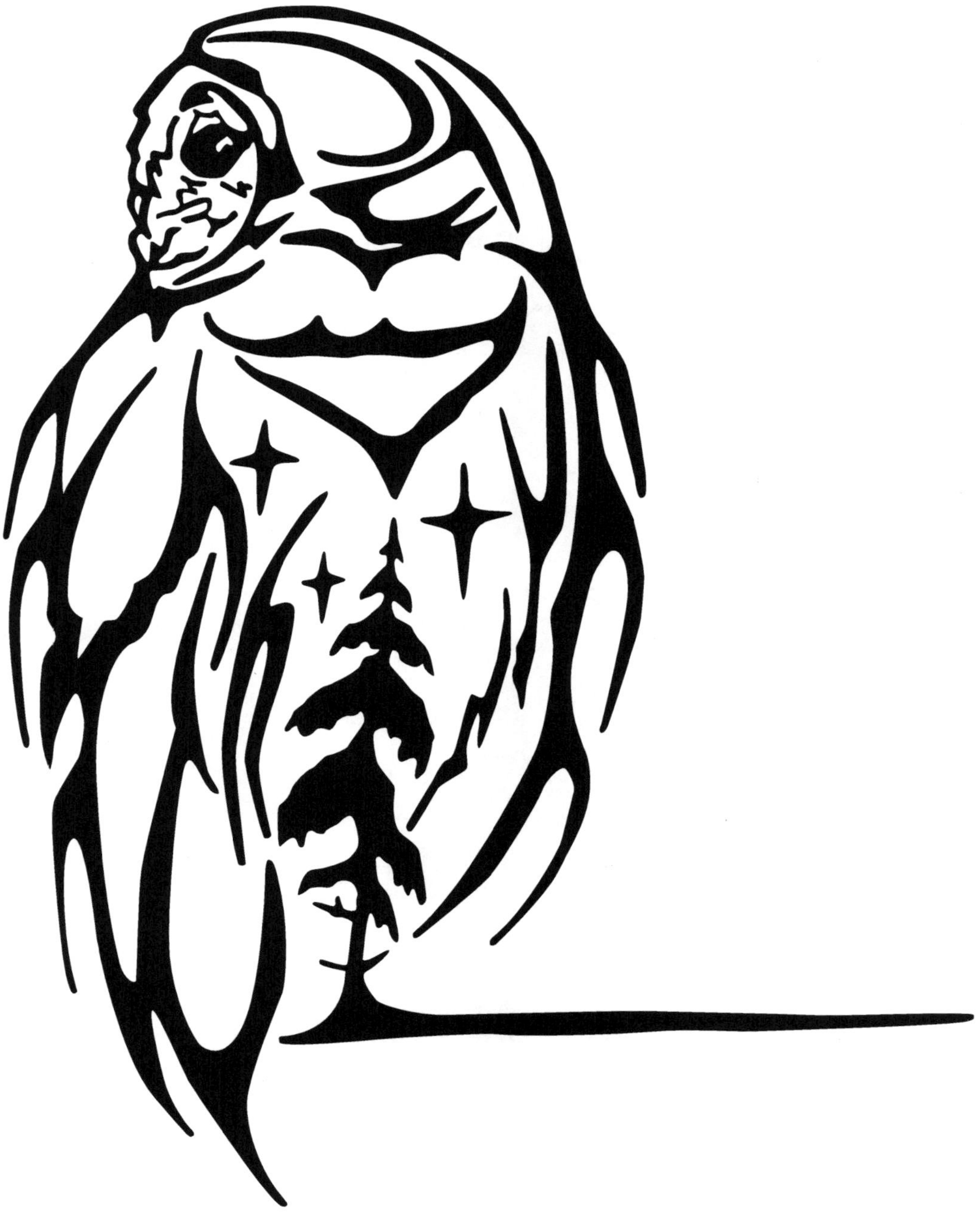

1 BEZHIK ONE

Bezhig Gookooko'oo ganawaabi gii-dibikak

One owl watching the night

2 NIIZH TWO

Niizh Migiziwag ishpi-babaamaashiwag

Two eagles high up in flight

3 NISWI THREE

Niswi Makwag bapaakwa'amoog

Three bears coming out of their den

4 NIIWIN FOUR

Niiwin Moozoog babaa-ganawaabiwag

Four moose looking to the horizon

5 NAANAN
FIVE

Naanan Waagoshag mamaandaagochigewag

Five foxes playing tricks

6 NINGODWAASWI
SIX

Ningodaaswi Amikwag asiginigaazowag

Six beavers gathering sticks

7 NIIZHWAASWI SEVEN

Niizhwaaswi Bizhiwag andawenjigewag gaa-bangishimong

Seven lynx hunting at sunset

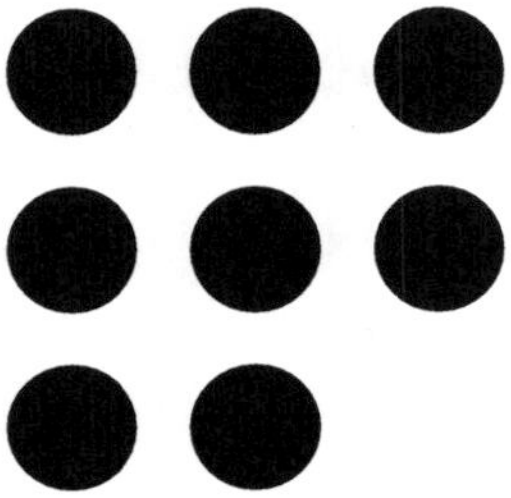

8 NISHWAASWI EIGHT

Nishwaaswi Asabikeshiikaadan ozhichigaadewan asab

Eight spider legs spinning a net

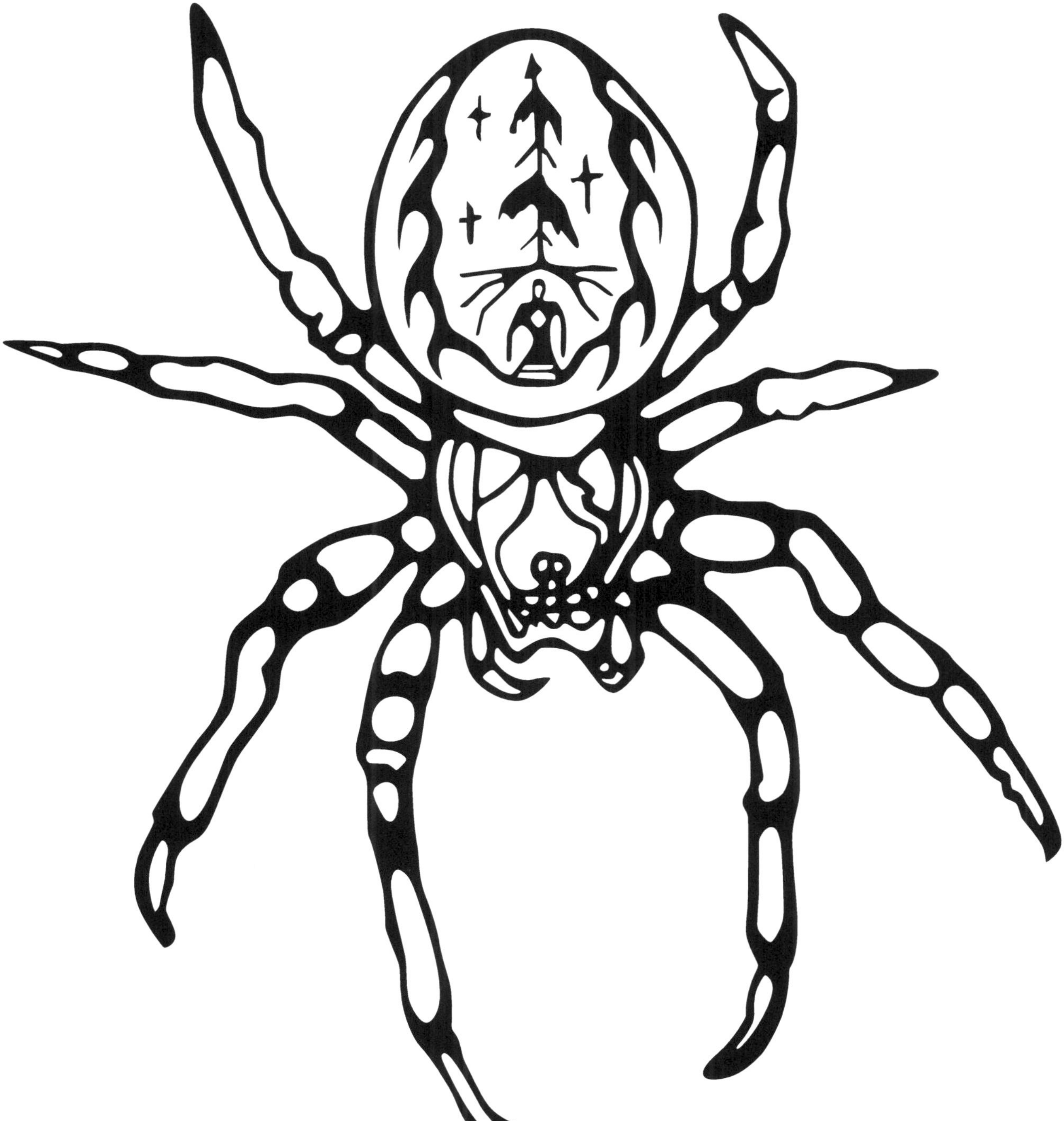

9 ZHAANGASWI NINE

Zhaangaswi Namebinag gii-siigwang

Nine sucker fish spawning in spring

10 MIDAASWI TEN

Midaaswi Zhiishiibag Bimaadigewag niisaajiwan

Ten little ducks swimming downstream

NOOKOMIS *GAYE* **NIIN**

nindibaajimomin gaye gaa-bawaajigeyaang

NOOKOMIS *AND* **ME**

sharing stories and dreams

What does *Connecting with Our First Family* Mean to You?

Indigenous peoples have held relationships with the land, animals, plants, waterways, and cosmos for time immemorial. Our health and wellbeing is tied to the wellness of all of creation. *Connecting with Our First Family* is not only about learning about animals, but also increasing our awareness of how we may walk alongside our animal brothers and sisters in a way that fosters sustainability, reciprocity, and respect for our shared mother, Earth.

How will you choose to honour this relationship?

Ideas for *Connecting with Our First Family:* With The Text

Ecosystem Mapping: Encourage students to think of the ideal ecosystem and environment. What do we need to live well? Invite students to draw ecosystem or community maps of their own ideal environment as well as the ideal environment for one of the animals. What are the differences and similarities? Is there a community we could map out that would have the ideal conditions for all of the animals and humans to live well in?

Drawing with animals: Pair with Connecting with Our First Family Colouring book or invite students to draw their own versions of the animals. Encourage students to think about the spirit and characteristics of the animals as well as their habitat. Invite students to share stories of their animals and how they might support those animals to thrive.

Animal stories: Spend some time reflecting on each animal image. What are some aspects you see in each image? What stories are being told through the image? What stories do you have about these animals?

Incorporating language into the classroom: Use the number translations into other aspects of the classroom. Have students number themselves in a circle reciting the numbers; count the numbers each day as you recite the date. Encourage students to research local Indigenous words for other things in the classrooms and label them to repeat.

Card Games: Pair with Connecting with Our First Family Learning with Numbers and/or Learning with Animals card decks. As you read the story, have students find the corresponding animal card. The student who finds the card first can be name the animal in English and in Anishinaabemowin

Ideas for *Connecting with Our First Family:* In The Community

Create an art project to showcase the importance of our animal and plant relations to our collective wellbeing

Visit a local Indigenous agency or friendship center to attend a workshop or community event

Invite Indigenous community members to share their own stories

Research the pre-colonial history of your community and talk about how the land has changed. What impact has this had on plants, animals, and people?

Organize a community cleanup to give back to the land you inhabit

Plant a peace garden to improve animal habitats in your area

Lead a nature walk and encourage students to think about how they feel in different spaces

Extend Your Learning: Numbers

1	Bezhig	One
2	Niizh	Two
3	Niswi	Three
4	Niiwin	Four
5	Naanan	Five
6	Ningodwaaswi	Six
7	Niizhwaaswi	Seven
8	Nishwaaswi	Eight
9	Zhaangaswi	Nine
10	Midaaswi	Ten
11	Ashi-bezhig	Eleven
12	Ashi-niizh	Twelve
13	Ashi-niswi	Thirteen
14	Ashi-niiwin	Fourteen

15	Ashi-naanan	Fifteen	**21**	Niizhtana ashi-bezhig	Twenty-one
16	Ashi-ningodwaaswi	Sixteen	**22**	Niizhtana ashi-niizh	Twenty-two
17	Ashi-niizhwaaswi	Seventeen	**23**	Niizhtana ashi-niswi	Twenty-three
18	Ashi-nishwaaswi	Eighteen	**24**	Niizhtana ashi-niiwin	Twenty-four
19	Ashi-zhaangaswi	Nineteen	**25**	Niizhtana ashi-naanan	Twenty-five
20	Niizhtana	Twenty			

Extend Your Learning: Animals

Recap the translations from the book as well as the others in the card deck.

	Makwa	Bear
	Waagosh	Fox
	Giingwa'aage	Wolverine
	Zhashagi	Blue Heron
	Marten	Waabizheshi
	Owl	Gookooko'oo
	Northern Pike	Ginoozhe
	Lynx	Bizhiw
	Sucker (Fish)	Namebin
	Mink	Zhaangweshi

	Waabooz	Rabbit
	Waawaashkeshiinh	Deer
	Nigig	Otter
	Diindiisi	Blue Jay
	Wazhashk	Muskrat
	Migizi	Eagle
	Mooz	Moose
	Amik	Beaver
	Ma'iingan	Wolf
	Zhigaag	Skunk
	Mashkode-bizhiki	Buffalo

	Aamoowadiswan	Bee Hive
	Maang	Loon
	Memengwaas	Moths
	Gayaashk	Seagul
	Miskwaadesi Gaye Mikinaak	Painted Turtles & Snapping Turtles
	Gaag	Porcupine
	Waawaabiganoojiinh	Mouse
	Nenookaasi	Hummingbird
	Zhiishiib	Duck
	Adikameg	White Fish

	Agoozomakakii	Tree Frog
	Nika	Goose
	Gaagaagi	Raven
	Adik	Caribou
	Asabikeshiinh	Spider
	Ozhaawahko-amogowesiimanidooshens	Caterpillar
	Ginegig	Snake
	Memengwaa	Butterfly
	Opichi	Robin
	Aanzonamegos	Rainbow Trout

Resources and Further Reading

OUR FIRST FAMILY

An initiative to support students and educators in the process of understanding and appreciating the diverse Indigenous languages, cultures and communities while strengthening connection to identity and relationships to the land.

www.ourfirstfamily.ca

CONNECTED NORTH PROGRAM

The Connected North program delivers interactive education services to schools in remote Northern Indigenous communities through high-definition two-way video communication and collaboration technology.

www.connectednorth.org

CONNECTED NORTH WELL-BEING FRAMEWORK

Developed as a learning resource and collaborative planning guide, This framework was developed by Connected North team members based on experiences and conversations with community members during visits and video conferencing sessions.

www.connectednorth.org/wellbeing

WHOSE LAND

A project bringing together land acknowledgements from across the country and an interactive map highlighting the traditional territories, treaties and Indigenous communities across Turtle Island.

www.whose.land

#RISINGYOUTH

The RisingYouth Community Service grants offer funding of $250, $750, or $1,500 for young people (15-30) proposing ideas to improve or address an issue in their communities.

www.risingyouth.ca

WE MATTER

An Indigneous-led non-profit commuted to Indigenous youth empowerment, hope and life promotion.

www.wemattercampaign.org

Made in United States
Troutdale, OR
11/03/2026

56844444R00027